Scan, A Serial Bus of N Bits Width Which Requires Bit Lane Deskew

Generic Treatment Of GSD Scan Testing

Edward Seymour

Scan, A Serial Bus of N Bits Width Which Requires Bit Lane Deskew

Generic Treatment Of GSD Scan Testing

Edward Seymour

Copyright 2016

Edward Seymour

ISBN

This text is dedicated to revealing a new paradigm on scan testing that can result in drastic reduction in test time with increased efficacy.

Table of Contents

Scan, A Serial Bus of N Bits Width Which Requires Bit Lane Deskew 1

Generic Treatment Of GSD Scan Testing......................1

Scan, A Serial Bus of N Bits Width Which Requires Bit Lane Deskew 2

Generic Treatment Of GSD Scan Testing......................2

Edward Seymour..2

About the author 6

Background 7

Crux Of My Idea...8

51 in 8

Say What?...9

Think For a Minute About Don't Care Shift Enable..........10

Simple Fix...11

Demonstrated in Every PowerPC chip In IBM, Apple, Motorola (1992-2004) 12

Supported Scan Rate For BIST of 5GHZ on Power7 Processor 13

Takes < .0001% area and Much Less Power...............14

25 Years and Products with 50 other companies.............15

Can Cite The Papers 16

People Mentored In This Include Master Inventors 17

Adament Supporters Of Latched Scan Enable 18

IBM Yorktown 19

IBM Almeden 20

IBM Zurich 21

IBM Yasu 22

IBM Haifa..23

IBM East Fishkill..24

IBM Burlington 25

IBM Rochester...26

IBM Raleigh 27

IBM Hursley Lab 28

IBM Austin 29

IBM Austin Lab..30

Group Bull – France and Italy 31

AMD 32
Purdue 33
University Of Illinois..34
Penn State University 35
University Of Minnesota 36
University Of Tennessee..37
Countless Others...38
    Honest Truth...39

About the author

I began my study of computer testing in Manassas Virgina when I ran a teaching program on University Campuses. Essentially I needed to affirm that chips were correctly produced. In this process I had the requirement to work with IBM inventors of Scan Methods. They had review rights on my texts for instructing undergraduate students in asic design and test. See my book on The-epitome-of-scan for details.

Background

A common misconception about scan testing is that it can be
"inserted" after all design is done. While true, in the factual sense,
if you want effective test time, some care should be taken in
timing of some connections, and/or setup of tester channels.

Crux Of My Idea

Assume you have a design with 500 pins, 100 of which are shared for scan testing. Depending on scan vendor, it will be 101 or maybe 102 pins total.

51 in

50

Conceptually, there is some magic which suggests that any random collection of chip pins, gathered in a group of 51 will be source synchronous, despite differing clock sources. Furthermore, it is current urban legend that a critical signal which determines the difference between shifting and not can be timed as a complete don't care for 2 or more clock cycles.

It was proven that shift enable IS a state signal and must be treated in the same fashion as any other clock gate in power management. If locally latched, it can be allowed any arbitrary insertion delay and will operate @ or above FUNCTIONAL FREQUENCY!

https://www.google.com/patents/US7895488

And countless others

Say What?

Imagine a flop comprised of Master and Slave

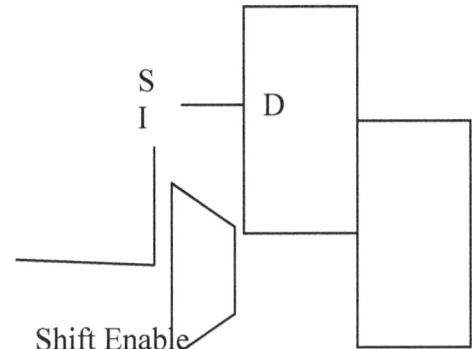

Think For a Minute About Don't Care Shift Enable

This implies

1. No Setup Check to Clock, Ever

2. No Hold Check To Clock, Ever

3. Allowed Skew between bits on the source syncronous serial bus of at least 2 clock cycles (more due to sum of errors)

Simple Fix

Locally latch shift enable and force setup and hold checks like data path. Insertion delay then sets dead period between shift and capture, capture and shift.

Demonstrated in Every PowerPC chip In IBM, Apple, Motorola (1992-2004)

Supported Scan Rate For BIST of 5GHZ on Power7 Processor

Takes < .0001% area and Much Less Power

25 Years and Products with 50 other companies

Can Cite The Papers

People Mentored In This Include Master Inventors

Adament Supporters Of Latched Scan Enable

IBM Yorktown

IBM Almeden

IBM Zurich

IBM Yasu

IBM Haifa

IBM East Fishkill

IBM Burlington

IBM Rochester

IBM Raleigh

IBM Hursley Lab

IBM Austin

IBM Austin Lab

Group Bull – France and Italy

AMD

Purdue

University Of Illinois

Penn State University

University Of Minnesota

University Of Tennessee

Countless Others

Honest Truth

Even implementation of a syncronous pipeline in shift enable is simple and cheap. ATPG tools were required 25 years back to support this in pattern generation.